# Aquaculture

# Introduction to Aquaculture For Small Farmers

# 2$^{nd}$ Edition

Kenn Christenson

## Copyright © 2015 Rach Helson LLC.
## All rights reserved.

In no way is it legal to reproduce, duplicate, or transmit any part of this document in either electronic means or in printed format. Recording of this publication is strictly prohibited and any storage of this document is not allowed unless with written permission from the publisher. All rights reserved.

By reading this document, the reader agrees that under no circumstances are we responsible for any losses, direct or indirect, which are incurred as a result of the use of information contained within this document, including, but not limited to, —errors, omissions, or inaccuracies.

The information provided herein is stated to be truthful and consistent, in that any liability, in terms of inattention or otherwise, by any usage or abuse of any policies, processes, or directions contained within is the solitary and utter responsibility of the recipient reader. Under no circumstances will any legal responsibility or blame be held against the publisher for any reparation, damages, or monetary loss due to the information herein, either directly or indirectly.

Respective authors own all copyrights not held by the publisher.

## Legal Notice:

This book is copyright protected. This is only for personal use. You cannot amend, distribute, sell, use, quote or paraphrase any part or the content within this book without the consent of the author or copyright owner. Legal action will be pursued if this is breached.

## Disclaimer Notice:

Please note the information contained within this document is for educational and entertainment purposes only. Every attempt has been made to provide accurate, up to date and reliable complete information. No warranties of any kind are expressed or implied. Readers acknowledge that the author is not engaging in the rendering of legal, financial, medical or professional advice.

# Table of Contents

Introduction . . . . . . . . . . . . . . . . . . . . . . . . . . . . . . . 1
Aquaculture – An Overview. . . . . . . . . . . . . . . . . . . . . . . .3
Some Pointers When Choosing the Type of Fish For Small Fish Farms . . . 15
Step-by-Step Guide on How to Set Up Your Own Fish Farm . . . . . . . . . 25
Some Fast Facts and Tips for Small Fish Farmers . . . . . . . . . . . . . . 41
Basic Aquaculture Tools and Equipment for Small Farmers . . . . . . . . . 45
Expansion for Your Fish Farm . . . . . . . . . . . . . . . . . . . . . . . 51
Marketing. . . . . . . . . . . . . . . . . . . . . . . . . . . . . . . . . 55
Conclusion . . . . . . . . . . . . . . . . . . . . . . . . . . . . . . . . 63
Photo Credits . . . . . . . . . . . . . . . . . . . . . . . . . . . . . . . 65
Other Books by Kenn Christenson . . . . . . . . . . . . . . . . . . . . . 67

*Aquaculture - Introduction to Aquaculture For Small Farmers*

# Introduction

Welcome to the world of Aquaculture for Small Farmers. This book contains steps and strategies on how to be successful in an aquaculture venture.

Aquaculture refers to the farming of aquatic creatures such as fish, mollusks and also plants. Just like sericulture and agriculture, aquaculture involves nurturing of natural elements and selling them for a price.

Aquaculture can seem like a daunting task for beginners but understanding it thoroughly will make it easier for you to adopt and practice.

This book provides small farmers and individuals with an extensive overview of aquaculture as well as some tips for setting up their fish farms. This book also features some practical tips for small fish farmers who are planning to grow fish outdoors or even in their own backyards. Also included here is a list of the basic supplies and equipment needed to start a successful aquaculture endeavor.

I want to thank you for purchasing this book, and I sincerely hope you enjoy it!

Kenn

# Aquaculture – An Overview

Consuming fish has become a popular trend in recent decades mainly because people are becoming more conscious of their health. Aquaculture farming, or simply fish farming, is the practice of producing fish as well as other crops that live in water. This technique has been around for over many centuries:

Since around 2500 BC, the ancient Chinese have been employing aquaculture. They held fish in synthetic lakes that formed after a river had flooded, and the waters subsided.

For over a thousand years, Hawaiian people have made use of aquaculture techniques, such as setting up fish ponds.

There is even proof that the early Romans cultivated fish.

Although in North America, fish farming is considered to be a very young industry. In contrast to more popular agricultural endeavors, a lot of people perceive the popularity of fish farming as a contemporary phenomenon.

Fish farming is now gaining popularity owing to the returns that it provides and also helps make families self-sufficient. The dependence on seas, lakes and natural water bodies to provide fish for consumption and ornamental purposes is slowly reducing and more and more people are taking to the art of aquaculture. The starting supplies required for aquaculture are relatively simple. All it takes to get started with fish farming are:

- A pH tester
- Feed
- Seed stock
- Aeration equipment
- Source of electricity
- A source of water
- A pond

## Types of Aquaculture Systems

### *Cage culture*

This type of fish farming technique provides the simplest method of cultivating fish if you have access to a pond. The only cost is for feed, fish and cage materials.

In this particular setup, a pen or cage made of rigid netting or plastic pipe is moored in any suitable body of water, such as millrace, stream, lake or pond, and loaded with fingerlings that are fed until the end of the growing season.

If you have a fish farm pond, setting up a floating cage will supply you with enough fish to feed your entire family throughout the year. The most common fish cultivated in cages is Channel catfish. Other good options include hybrid striped bass, salmon, trout and tilapia.

The biggest advantage of this type is that you will not have to shell out a lot of money. You will already have a ready water source available for you to start your farming, and this will allow you to save time as well. The disadvantage, however, is that, there have been several cases of the system failing to keep the fish inside the pond, and if yours escape, then you might have to suffer losses. There is also the danger of disease control as the water will be too vast for you to be able to control.

## *Flow Through*

This type of fish farming redirects a flowing source of cold water such as a river, spring or a stream into "raceways" that will hold the fish. Even a small amount of water can establish a more productive aquaculture system than a closed setup. With just a few gallons of stream water, you will be able to cultivate trout all year round.

Flow-through setups are very simple and are relatively inexpensive unless you don't have access to any natural source of flowing water. It is also essential to note that this particular setup is subject to regulations concerning diversion, as well as the use of natural water resources.

The advantage of this type is that, you will need only a little water, but it needs to be consistent and free flowing. As was mentioned earlier, you will be able to breed fish all through the year and so, you can look at this as an option for consistent farming. However, the biggest disadvantage is that, you will have to clean the raceway regularly in order to get rid of ammonia and other wastes from the water. This can be quite a tedious task, and you will have to make use of pumps at the bottom to drain away the dirty water. For this, you will have to avail permission from all the appropriate authorities in order to not get into trouble of any sort.

## *Greenhouse Aquaponics*

This technique uses different types of plants instead of filters to enhance the quality of the water. In addition to tilapia, hybrid striped bass, catfish or trout, you can also cultivate various types of vegetables such as cucumbers, tomatoes, and lettuce, as you would in traditional hydroponics. However, since fish thrive in water, the nutrient sources for greenhouse Aquaponics must be entirely organic; no fungicides, insecticides or herbicides shall be utilized. This can make it a little difficult to maintain the plants.

This technique is considered the most complicated fish farming method, needing a high level of management as well as other components such as aerator and water pump. The costs will only go up, and it might Backyard aqua-culturists usually combine aquaponics with a simple re-circulating set-up.

## *Home Recirculating*

This technique is the answer for would-be fish farmers whose only source of water is a garden hose. The most excellent method to put-up this miniature fish farm is to utilize an above-ground swimming pool in a basement, garage or backyard. The cost of such swimming pools would differ depending on size. They are heavy-duty, with only the vinyl liner requiring replacement after a couple of years. This technique is discussed in detail in the next chapter.

## The Essential Qualities of a Successful Fish Farmer

Fish farming is likely to be more appealing for the patient person. Males should not be discouraged from venturing into fish farming, although according to the International Food and Agriculture Organization, females tend to be the most successful fish farmers. Different studies reveal that women are more likely to be methodical and meticulous. Women will have the patience to sit down and sort out issues much more effectively than men. However, if a couple decides to take it up, then they will surely be quite successful.

Apart from patience, it is also important for the fish farmer to be well read. This means that the farmer needs to do some research and understand every little aspect of fish farming in order to be successful.

Be mindful of the pond: It is essential that the fish farmer understand the pond since, in essence, it is not the animals that are managed. What is actually managed is the environment where the fish lives in, which is the pond.

It takes an individual who is in touch with the pond to know what it is doing and what it is going to do before it even occurs.

## Facing the Challenges of Fish Farming

In contrast to other types of conventional crops, the investment returns can be a lot higher on a per acre basis. In addition to this, most aquaculture crops will only need a growing time of about 4 months, lessening the time when cash flow could be an issue. Some aquaculturists don't survive this initial period and decide to stop the practice as the profit margin will start to deteriorate if the costs keep rising.

Searching for a Knowledgeable Assistance – one of the major downsides of endeavoring in fish farming is that the knowledge base is a lot smaller as compared to those of traditional crops. Still, when difficulties arise, it can be very challenging to find others with the expertise to help you out. There are many farmers who give up on farming only because they will not have valid solutions for their problems. They will start to have these problems accumulate and this will only cause them losses. There are fish farming communities that might help provide solutions but these might not be located in the area where you have your farm and, therefore, cause you to have problems and not find solutions for them.

Trying to maintain large varieties can pose a big problem. Not all fish will adapt to the same kind of environment, and this will force the farmer to have just one variety. But a single variety will not bring in good business and trying to introduce different species will only add to the cost. Many farmers try and club different fish that can live in the same type of environment and end up overcrowding. This can prove to be disastrous as not only will the quality of the pool go down, but the fish will turn out to be unhealthy as well.

The problem does not end there; mere variety will not cut it, and the farmers will be required to grow exotic fish. But growing exotic fish will require permission, which is not easy to avail. There have been several cases of mutations, which only further endangers the fish communities. They will also not be edible, and this will limit the market for these farmers. They will have to choose between ornamental or edible and not be able to have both together.

There are several cases of these toxic and hybrid fish escaping from their cages and mixing with wild fish and finding their way on plates. This has raised a cause for concern amongst the people who consume fish regularly, and there are several protests against aquaculture. This has again caused a lot of losses amongst fish farmers.

Another problem with aquaculture is that, the fish are fed natural food, which is causing for food resources to deplete. Most fish food is made from other sea creatures, and this is causing a reduction in the wild waters. It takes a lot more creatures to prepare the food as opposed to what these fish would consume if they were to live in the wild.

The overall cost of starting a farm is also quite high. Once set up, the maintenance costs can also rise, which can deter people from taking up the practice.

Fish farming is an emerging market – although by far, the biggest challenge for this young industry, surely for the small farmer, is the absence of market structure. The majority of aquaculture producers have to deal with marketing their very own products firsthand, all the time working against the clock with extremely perishable goods that, similar to fruits, have a very short shelf life.

You will have to do everything from scratch if you wish to market your business. There will be no established route for you to take and will have to conduct the research, invest money, advertise, etc. More on this is discussed in the chapters to come.

## Threats to Your Stock

### Few predators; although watch out for stress

Minimal external danger exists. Snapping turtles are not a problem; raccoons are thwarted by the pond's depth, and wading birds are unable to wade far. In the case of freshwater prawns, however, they get stressed out too easily. This can cause them to develop swim bladder diseases as also end up producing weak spawn. Freshwater shrimps are severely territorial in nature and turn cannibalistic under duress. There have been several cases of 10 or more shrimp being put into a pond but only a couple being alive in a few days' time. That is how risky shrimp farming can be and so, many aquaculturists try and steer clear of it. Poor pH balance and lack of oxygen can cause a whole pond of fish to die, which can be quite disastrous.

## Diseases

There are usually only a few basic diseases that apply to fish, typically due to poor care. Many times, we tend to assume that fish are strong enough to survive anything as they live in the water and will have a great deal of immunity. But the reality is just the opposite. Fish will only be hardy if they are left to survive in their natural habitat. Raising them artificially and feeding them artificial food will only make them weak, and poor conditions will cause them to develop diseases. It is quite easy for these diseases to travel from one fish to the other and so, it becomes important to get rid of the diseased fish as soon as possible.

Artificial environments as an option – there are certain commercial producers who have set up artificial environments with hydro-technical and recirculation systems; in general, these strategies necessitate a more advanced experience level, as well as more technical and energy-efficient setups. You may be interested in this type of set-up to support the life of your fish.

## Unpredictable environment

Just as like any other types of crops, a more common threat to an aquaculture harvest is nature's unpredictability. Shrimp and fish are cold-blooded creatures; inevitably they are entirely captives of their environment. So the number one factor in cultivating such food sources is temperature.

Temperature matters almost more than any other factor. There are various categories, including cold, cool, warm and tropical water, and the climate dictates the type of species that is suitable and possible to cultivate.

Hot spells and cold snaps can have a distressing effect on pond life. Imagine a few hundred of your fish all dying just because there was a slight rise or fall in the temperature. It becomes all the more important to control the environment that prevails outside the tank and be able to give the fish the right sort of atmosphere to thrive in.

**Temperature**

Power failure is among most threatening issues. After all, pumps are sometimes needed to supply sufficient amounts of oxygen. The power costs for fish farming will be very high. If you are unable to pay the bill for it, then you might be in trouble. Some fish farmers try and harness solar power for their farms as it is sustainable and renewable source and also much cheaper. But the initial set up cost might be a bit too much, which is why only a few farmers adopt the concept.

# Conservation and Environmental Advantages

Fish farming perhaps has the most excellent potential for relieving pressure on endangered wild populations and letting the conservation efforts on those populations be successful. It is easy to isolate the endangered species' and take care of them to help them multiply rapidly. In fact, several rare species have benefitted thanks to fish farming.

The ecological footprint of fish farming is relatively small. Everything has a cost; although compared to wild fisheries, fish farming is a relatively low impact enterprise.

You might want to check with your government for any schemes that they might have to offer as fish farming is a good way to contribute to the environment. You might be able to avail loans at discounted rates of interest and also have free or subsidized assistance meted out to you as a means of government aid.

# Some Pointers When Choosing the Type of Fish For Small Fish Farms

Fish serve as the powerhouse of an aquaculture system as they supply the nutrients for the plants. If you are raising edible fish, then they will also supply protein for you. Raising fish could be somewhat daunting for some people, particularly those individuals who do not have any prior experience; however, this should not make you feel discouraged. Raising fish in a backyard fish farm is actually a lot simpler than keeping fish in an aquarium. As long as you follow the simple tips, then raising fish from fingerlings to ready-to-eat sizes can actually be easy.

There are lots of fish species that may be used in an aquaculture system, depending on the available supplies, as well as your local climates. Winter climates could allow for raising Rainbow Trout while warmer climates can allow for raising "warmer species" such as Barramundi. There are several all-year-round fishes that you can grow; however, these fishes usually take a long time to grow. If you are located in colder regions, then you should be looking at raising Trout all year round or probably another locally-produced fish species. In warmer regions, people usually grow Jade

Perch or Barramundi throughout the entire year. In the majority of areas throughout the world, Tilapia is the fish species of choice.

When deciding on the best fish species to grow, you must take into consideration some important factors. The most important of which is what you want to derive from your system. If you do not wish to consume your fish, then you probably do not want to raise edible fish. Otherwise, you may wish to grow edible fishes that can thrive year-round in your area. The second most crucial consideration is what's available. After all, you need to purchase fish to stock up your aquaculture system. To help you in determining the perfect fish species to grow in your small fish farm, below is a list of valuable aquaculture fish and some details about them:

## Trout

Trout are an excellent fish species for most aquaculture systems where the water temperature is a little cooler. Trout thrive very well in water temperatures of 10 to 20 degrees Celsius (50-68 degrees Fahrenheit). Trout have very fast growth rates as well as excellent food conversion ratios. You will be able to break into profits in no time as they will be ready to be sold.

## Tilapia

Tilapia are considered the second most cultured fish in the entire world. Tilapia are also extremely popular in most aquaculture systems for a lot of reasons, such as the following:

They are omnivorous and are good eaters so you will have a clean tank

They grow very fast which means that you will be able to sell your fish quite fast

They can withstand very poor water conditions and so, you will not have to worry too much about their environment

They are easy to breed and will only need a little cover

The only drawback for certain people will be that Tilapia require warmer temperatures. If you reside in colder places, then you are far better off raising a fish species that will do well in your temperature range, rather than try to heat the water. In a lot of areas, tilapia is considered as pests.

## Silver Perch

These fish are excellent all-year-round species that are native to Australia. Silver Perch thrive well under different conditions. They are omnivorous and will happily consume green scraps such as Azolla and Duckweed. Silver Perch thrive within a wide temperature range, although they do not grow as fast as most other fishes. Fingerlings take about 12 to 18 months to grow into plate size. You will not have a problem breeding them in large numbers as they are used to being that way in the wild. They will make for ornamental fish, and you will be able to supply them as and when needed. They are extremely hardy and will be able to undertake any water condition. However, you must not ignore them as dirty water conditions can cause diseases and also a low reproduction rate. They are extensively eaten in Asia, and you will find it easy to sell these to Asian restaurants and also to Asian supermarkets.

## Murray Cod

These fish are also native to Australia and are commonly known to grow to huge sizes in their native habitats. Murray cod are raised in recirculating fish farming systems. Murray cod grow fast and are an excellent eating fish. One of the drawbacks is that Murray cod should be maintained at high stocking densities and must be well fed. Otherwise, they tend to cannibalize each other. They are extremely aggressive and highly temperamental. If you wish to keep them with other fish, it is best that you create a barrier between them and the other fish. They love to move around so make sure there is enough place for them to do so. They have to be kept in fresh water and cannot handle salinity. It will be a bad idea to place them with shrimps as they will not be able to adjust. Most cod are susceptible to pathogens, and it becomes important to check them for diseases from time to time.

## Koi

Koi are another species of carp and are very popular among Asian communities. Koi are usually found in huge ornamental ponds. For individuals who love Koi, an aquaculture system is an excellent idea for stocking the fish. The fish are said to live for more than 60 years, and this feature makes them some of the fastest selling fish in the world. Koi are also considered to be quite lucky and having one will only bring you luck. But they might need a pool of their own if you wish to use them for your purpose. There are several varieties of Koi to choose from and you can choose between orange- white – black or black- white – silver, etc. all varieties of Koi are beautiful and you will have a great time breeding and selling them.

## Carp

There are a lot of different species of carp that may be very well suited to aquaculture systems due to their ability to easily adapt in a lot of areas around the world, their tough nature and their reproductive abilities. However, carp have become damaging to native waterways. In the majority of western countries, carp also aren't well liked as food, although they are still the most widely cultured fish around the world and are cultivated throughout the entire Asian region. Carp will easily adjust to your pool, and you will have no problems in getting them to reproduce. There are different shades that you can choose from, and they can also be raised as ornamental fish. It is believed that they are tough to catch in the wild as they can be quite quick, and this only causes them to be highly priced. The mirror carp is a variety that is said to be low maintenance and tastes the best. You will be able to make a lot of profit if you decide to farm carp, and it is best that you include them in your farm.

## Barramundi

Barramundi are usually cultivated in aquaculture systems during the warmer months of the year. The majority of Barramundi growers purchase fairly mature stock so that they will be able to harvest bigger fish at the end of the growing season. When grown in an aquaculture system, Barramundi has an excellent clean, crisp taste. At the end of the growing season, they provide a decent harvest. Barramundi are considered one of the most majestic species of edible fish. They have white flaky flesh that is said to cook very well. They are generally served with lemon wedges as it offsets their natural taste and are a staple of the Thai cuisine.

Barramundi are also ornamental fish and so, you will be able to sell them as both. You can keep your barramundi with other species as long as they are not smaller than them. There are several varieties that you can choose from, but the most common type is the all silver or the all silver with black spots.

## Shrimp

Some small fish farmers find success growing freshwater shrimp. But remember that your shrimp are extremely high maintenance. The water needs to be saline, and this means that no other fish can be put in with them. But there is an advantage to growing shrimp in a separate tank as most places that sell shrimp will look for farm grown ones. This is because a majority of the commercially grown shrimp are said to be quite adulterated and through testing, it was found that they can contain dangerous chemicals as also some antibiotics in their bodies. This has caused people to reject the commercially grown ones and also the ones that are imported. Another report also claimed that most shrimp packing plants are quite filthy and so, packing them yourself will give you a further advantage. It might cost you a little money to set up the ideal farm for the fish but once you do, you will be able to attain good levels of profits.

## Goldfish

While some individuals may categorize these with the carp, it is important to cover goldfish separately. Goldfish are usually sold as at local pet shops. Goldfish are generally tough and make an excellent addition to an aquaculture system. In most areas, goldfish are capable of breeding in a tank; however, they require plant cover within the tank in order to multiply. You can add in natural plants or have plastic ones. You can also have rocks and prepare a cave for your goldfish. Sometimes, extremes in conditions can cause your fish to undertake stress, which can be a bad thing. Make sure that the fish are in an optimal environment and are not bothered by any environmental issues.

Apart from aquarists, goldfish are also popular amongst anglers and fisherman as they are cheap and are loved by other fish. In fact, several farmers grow goldfish to be sold as bait and so, your ornamental fish might be of more value than you think. Goldfish should not be clubbed with other types of fish unless they are more of less of the same size. They will eat fish that are smaller than them and be eaten by fish that are bigger. There are a lot of colors that these fish are available in and you can choose between completely gold to silver with orange spots and also the black variety that are said to bring in luck when placed with the other fish.

## Channel catfish

The channel catfish is a great fish for you to grow in your backyard. The channel catfish is not reputed for being an edible fish as they are known to be scavengers. But these fish can be extremely tasty and is extensively eaten in the south. These fish can also be bred as ornamental fish as they look quite good, and their two prominent whiskers makes them quite unique. This fish is popular with fish farmers, and it is believed that nearly 80% of them are bred in farms. The catfish is easy to maintain, and you will have the advantage of having a scavenger in your tank. This will ensure that the tank remains clean and will help reduce your work. There are different varieties available for this fish, and you can choose between the all silver one to the one with polka dots on its body. These fish go well with other fish and also with their type. That will allow you to mix them with other fish, and it will be possible for you to keep them with smaller varieties as well.

## Eel

If you are interested in having a little variety, then you can also give the eel a try. Baby eels can be quite the site and are some of the most majestic creatures. Eels are considered to be a delicacy and can be sold to restaurants. They are also quite popular as ornamental fish and so, you can grow them as both. But it is not advisable to have your eels in a cage system as they will easily slip and escape from in between the mesh. It is best to have them in the raceway or your backyard pool. They can be a little high maintenance, but they will surely fetch you a good price.

## Bullheads

The bullheads resemble the catfish but are much more colorful. You can choose between the yellow, brown or black variety, and each one can be quite a delight. Bullheads taste great but are not eaten enough. That is owing to a lack of knowledge and more importantly, its unavailability. Bullheads are easy to breed in your backyard and will only need some cover. You can try giving them rocks and also make a cave for them. They also make for ornamental fish so you will be able to market them as both.

## Bluegill

The bluegill is a fish that will be ideal for your pond farm. It is the best-suited sunfish for aquaculture, and you will be able to have a good business. These fish are popular for being hybrids and can be bred as ornamental fish. But they are expensive and small farmers might not be able to afford the hybrid varieties.

These form the different fish that you can consider for your farm but remember that there are several more, and you can choose the one that will fit into your budget.

# Step-by-Step Guide on How to Set Up Your Own Fish Farm

Starting out with your own aquaculture endeavor might sound like a daunting task at first, but believe me, it is one of the easiest things that you can do. Here, we look at the things that you must consider while starting out.

## Planning

Planning out your aquaculture project will help you move in the right direction. You will have to sit down and plan every small detail beforehand. Some people decide to go with the flow and get started without a plan in mind. This will only cause them to face problems later, as unforeseen issues will crop up. So no matter how boring or tedious it sounds to sit down and sketch out a fool-proof plan for yourself, you simply must do it, in order to avoid problems later and also make it easy for yourself.

## Deciding

Deciding on whether you wish to make it a small pond culture or a big scale one, deciding on having a separate pond for each fish, deciding on having or not having plants, deciding on having shrimps and other creatures etc. etc. are all important choices that you will have to make for yourself. You can consult with a friend or a relative and then make your decisions. Just like planning being an important phase, decision-making is also an important part of your starting process.

## Budget

Decide on the budget that you wish to employ for your project. Again, not deciding on it and simply jumping in will cause you to splurge more than necessary. You must aim at minimizing all your costs and set a budget that is ideally less than what you think the overall cost will be. That way, even if you overshoot your budget, you will remain in the confines of your estimation. Having a fixed budget will also help you make arrangement for the money, and you will not have to scamper around later, in case you need some extra dollars.

## Assistance

Once you decide on the previous three, you can enlist the help of a friend, a sibling or a relative to get started. Even if you wish to be the sole proprietor of your business, you will need a little help while you set it up. You can promise them a reward or can also ask them to join hands with you if they are also looking to invest in a business. This will take off some responsibility from your shoulders, and once people start to invest, they will begin to take a keen interest in the affairs of the business.

## Know how

General know how of aquaculture will be an important asset for you. Before you start out, you must brief yourself on the various aspects that will be involved and for this purpose; you can do a thorough research on it on the internet or buy yourself a beginner's guide book. This book will help you in a big way no doubt, but the more that you will read, the more the expertise you will gain and the better the outcome that you will have. So start immediately as you can amass a lot more knowledge on the topic.

## Guidance

You can avail the guidance of a professional aquaculturist and visit his or her farm to have a firsthand look at what it takes to run an aquaculture farm. You will have an idea of the various things that you must do to have a smooth and successful farm as also the various arrangements that you will have to make for your fish. Remember that advice from an experienced culturist will be much more valuable than what you will read on the internet. They will give away important secrets that will make it easy for you to establish your own farm as also help you get started on the right foot. Don't refrain from asking them any doubts that you have about the farm and make sure you persist at it until you have a satisfactory answer.

## Purchasing equipment

The next thing to do is to purchase the equipment for the farm. The equipment includes tanks, nets, motors, plants, rocks, etc. You must always buy the best quality equipment as it is important for them to last long. You can find a good dealer in your locality and look at their testimonials online. If you have a friend or acquaintance, then that will also help you greatly as you will avail better discounts. If you are buying everything in bulk, you must ask for a discounted price, and it is best that you look at two or more vendors to help find the best one. This topic is discussed in detail in chapter 5 of this book.

## Purchasing fish

The next step is to purchase the fish themselves. When you purchase fish, you must buy them from the right vendor. You have to consider the cleanliness standards, his testimonials, and the general health of the fish. Remember that it is easy to assess the health of fish just by looking at them, so make sure you have a look at the fish at the farm where you purchase them from. If you are buying the spawn, then ask for the best quality ones. You can try and purchase different varieties from different sellers as this will help you have a mixture of varieties.

These will help you get started with your aquaculture. Remember that it is best that you follow this particular order as it will be easier for you to go about it in an organized fashion.

## Location

The next big thing that you must consider is the location of the aquaculture farm. The right location will help you run a successful farm as also allow you to save on costs. You will also not have to worry too much about the maintenance as the right location will be conducive for your fish. Whether you plan on having a backyard pool or invest in aquaponics, you must consider the following aspects.

Here are the various things that you will have to consider when choosing the right location for your aquaculture farm.

## Water supply

The very first thing to consider is the water supply that will be available at the site. It is obvious that water is the most important resource that you will need when you decide to start an aquaculture farm. The water needs to be in abundance, and you must not choose a site where there can be the danger of drought.

Remember that you will have to have a large water source close to your farm. This can be a lake, a reservoir, a river, etc. You can also have a well that you can dig and harness the natural underground water. You must check the drainage policy of the lakes as you might end up getting abundant water that might cause your farm to overflow. This can, especially, be apparent during the rainy season and you must check how well the water body drains.

You must also consider the rain water and decide on whether you will use the water for your farm or if you need to cover it up to prevent the water from being added to your pond.

## Water quality

The next thing to consider is the water quality of the pond. It is important that you check the quality of water as fish will need a certain composition in order to last longer. This is to be done by collecting samples from the source and testing each sample to assess whether the water is good enough for your fish. It is also important to collect it over a few months' time as the composition of the water changes from time to time. You must look into the history of the water source and make sure that there were no incidents of contamination that were prominent.

Once you collect the water, you must check it for its physical properties such as its general temperature, its color, the particles present it, its viscosity, its odor, etc.

You must then move on to its biochemical nature and look into its composition, its pH level, salinity, alkalinity, the presence of any chemicals, pollutants, etc.

You must then check for parasites, their presence, their eggs, any traces of microorganisms, bacteria, fungus, etc.

Once you think that the water is suitable for your type of fish, you can decide to use it for your farm.

## Climate

The climate of the location will greatly determine how successful or unsuccessful your endeavor will be. Many people do not realize that in order for fish to remain healthy, they also need to consider the external environment, i.e. the environment that exists outside the tank or pond.

This is important as the environment will have a bearing on the water and determine its very nature. For example, if there is a very warm climate, then the water will evaporate quickly and so, it will be important to refill the tanks or ponds daily. If the temperature is too cold, then there will be the danger of ice formation.

So it is important to understand the type of weather that prevails around the locality in order to choose the best spot. Here are some things that you must analyze:

- Average monthly temperature
- Average monthly rainfall
- Average monthly evaporation
- Average monthly humidity
- Average monthly sunshine
- Average monthly wind speed and direction

You must collect, at least, 6-month-old records in order to have a clear idea of the type of weather that exists over the plot of land where you wish to establish your fish farm.

All these will help you understand the type of weather that generally prevails over the particular land, and this will allow you to make your decision. You can avail these from the concerned authorities and be rest assured.

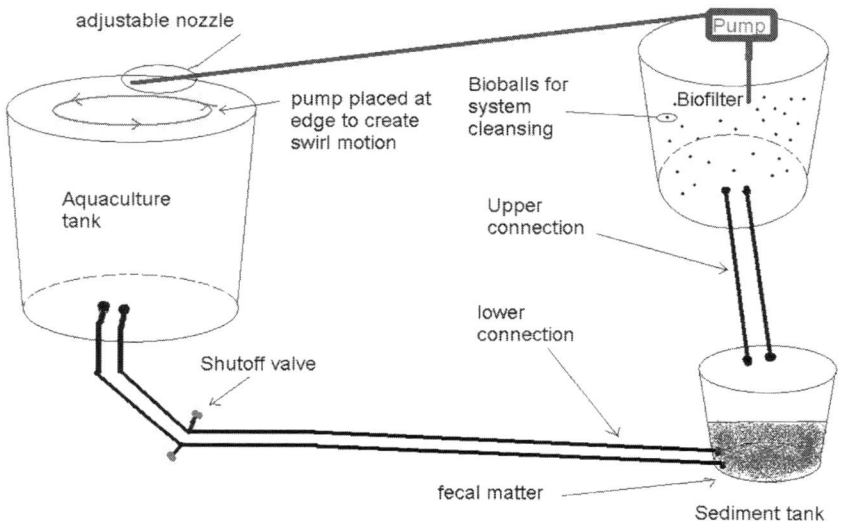

## Permissions

The next thing you must consider is the permissions that you must take from all the appropriate authorities in order to source the water from the lake, pond, and river, etc. You will have to ask for a monthly allowance of the water into your farm and also ask for it to be cut off for a little time as you might have to perform cleaning or undertake other activities for your farm. You must not get into trouble later so make sure you take all due permission before starting your farm. If you have to take the permission of someone else, like maybe neighboring farmers, then it is best that you do that as well, as you do not want to get into trouble anytime later.

## Soil characteristics

Soil characteristics refer to the condition of the soil where you wish to lay down the pond or fish farm. Sometimes, the soil might be loose, and it might not hold the pond well. This might cause you to suffer losses and so, it is important that you check the soil thoroughly before deciding to set up your farm.

Although a visual examination will be good enough, it would also be ideal for you to pick up a little and have it tested just to be double sure.

All these form the various aspects that you must consider when you decide to start your aquaculture farm. You will have to put in a little effort in the beginning and make sure that you take into consideration every little aspect of the location. You will only invest in your fish farm once and so, it is important that you employ all your efforts into choosing the best and most ideal spot for it.

## Backyard farming

Why buy costly, imported farm-raised fish when you can raise and catch your own right in the comfort of your own home? Prudent homeowners can effortlessly cultivate fish in their own backyards, making huge savings in their food budget. If you can maintain an aquarium, then you can probably maintain a fish farm.

With the increasing popularity of growing vegetables at home, backyard fish farming is also emerging to be another common means of boosting food supplies. Begin small by building your own backyard fish farm. Once you are adept, you can also expand and turn it into a business. More on that is discussed in the following chapter.

To set-up a fish farm in your own backyard, you will need the following things:

- Grass clippings or commercial fish food and algae
- Dip net or dragnet
- Charcoal filter for chlorinated water
- Water testing kit
- Sump pump or aerator
- Wading pool

Procedure:

- Put a 12x12 foot above-ground pool or a fish tank in a level area close to a power outlet. If there is no outlet, then you must have one installed as it will be mandatory to plug in motors and other equipment in order for your fish to thrive. Remember that the pool needs to be made of metal or sturdy plastic as it will be used to hold quite a bit of water. You must buy a good quality one as it will be a onetime investment. If you are up for it, then you can dig a little and then place the tank as it will provide it a bit more rigidity. You can have someone help you do it and it is ideal to first wet the surface of the soil and then dig or simply use a crane to make a large enough hole in the ground for your pool to fit in. Once you place the tank firmly in, you can cover the sides back with the mud that you dug, in order to seal it in better. If you are placing your pool on concrete, then you can avoid this step.

- The next step is for you to fill the pool with water. You must have a water source close by, such as a tap and you can use a large hose/pipe to fill it up. Load the tank with warm, filtered water. Depending on the type of fish that you would like to grow, the temperature you need may vary; although 75 degrees Fahrenheit is an average comfortable temperature for a lot of fishes. You will have to fit in a thermometer to keep an eye on the temperature and most thermometers are attached to heaters, which will automatically heat up the water in case it starts to get too cold inside the tank. If you have chlorinated water, run the water through a charcoal filter. Make sure that the pH level of the water is about 6.5 to 7. If the water is too alkaline, add gypsum; if it is too acidic, and then add lime. The water must also be tested for heavy metals as this can kill the fish. If there are heavy metals detected, set-up a filter to lessen the levels. Set up an aerator and make sure that it is properly working.

- You are now almost ready to have your fish in your pool, except that the water needs to get ready to host the fish. Have the filters, water and pool running for about 2 weeks prior to adding the fishes. This will help balance out the bacteria levels and condition the water. Some people make the mistake of not showing any patience and end up adding the fish without waiting for the pool to condition. Remember that there needs to be bacteria in the water that will help in converting the nitrates, which is essential for your fish's survival. You will know when the time is right for you to put the fish in the tank when a thin layer of mold will be visible all over the tank surface. If this layer is still not formed, then you can wait for 2 more days.

- The next step is to buy the fish. Remember to buy them from the best dealers as it is important to start with healthy fish. If you get diseased ones, then not only will you endanger the other fish but also have a pool that is contaminated. So look for trustworthy vendors and buy from. Another point to note is that, the dealer should be located close to your farm. Most fish do not handle transit well, and if they are clubbed in a single bag, then they will hate it even more. Fish are solitary creatures and so, you might have to separate each as much as possible. If the fish are delivered in plastic bags, put the bags with the fish inside in the tank to help them easily adapt to the change in temperature. After a couple of hours, you may start releasing the fish into the water. Remember to add in the water that came with the fish. This water will be familiar to them and adding it in will only make them feel at home. Don't leave any of the plastic bags inside the pool as this can be bad for them. You might want to keep an eye on them for some time as some fish might need to be isolated from the rest for a little time.

- For optimum outcomes, use or build your own floating feed rack to nourish the fish with minimal wastes. Alternatively, you may also feed the fishes by placing the food on top of the water. Fish feeds that are available commercially are customized to the particular requirements of certain fish breeds, such as Carp and Tilapia. Still, most fish may thrive on grass clippings and algae along with small amounts of animal manure. Feed your fish at about the same time each day and increase their feed as they grow. The fish must be fed about 3 percent of their weight every day.

- One thing to remember is to never overfeed your fish. Initially, you might have a problem understanding how much needs to be fed. For this, you must ask the vendor and stick to the amount that he or she tells you. It is easy to get carried away and over feed your fish, but this will not only affect them but also cause your pool to go dirty in no time.

- When harvesting your fishes, make use of a drag net to get multiple fish at a time; use a dip net to harvest a single fish. But some fish will start to panic if you use the single net method and they might start to stress out. In such cases, it will be important for you to release the fish back into the water and then remove it along with some other fish and separate it later.

## Some tips and warnings:

This is a very important point, so I wish to mention it here, once more. Keep from over feeding the fish as this can lead to problems with the water quality. It can also make the fish sick.

- Buy quality food from reliable sellers and don't try feeding them human food as that might make them sick.
- Use a fountain or any other device to ensure that the water is kept aerated.
- Check the temperature from time to time to make sure that the fish are comfortable.
- It is extremely important to remove dead fish out as soon as possible as the other fish might start to feed on it, and this can cause the pool to get dirty.
- Clean your pool at least once a week by removing out about 1/3 rds the water and adding in fresh water.
- Do not add in conditioners if you don't know the proportions. You might end up poisoning the fish.
- If you wish to add decorations, make sure that they are aquarium safe and clean them thoroughly before adding them in.

- Your fish will need some place to hide and so, you will have to place a few rocks and try and make a small cave for them.

# Some Fast Facts and Tips for Small Fish Farmers

While there are mixed opinions about aquaculture or fish farming, you can be sure of making considerable profits out of it. Despite some difficulties faced, the margin of benefit will remain unwavering. Below, you'll discover some facts and tips that you have to know before you indulge yourself in the world of aquaculture.

## Facts and Tips on Aquaculture for Small Farmers

- According to the United Nations, aquaculture or fish farming provides a considerable increase in the rate of employment all over the world. As a matter of fact, the increase in employment rate brought about by fish farming is expected to be doubled by the year 2030. This is good news as you will be helping in contributing towards providing employment opportunities to people, and this will, in turn, help increase your business.
- The reason behind this surge is the tremendous increase in worldwide population as well as the increasing food requirements. More and more people are taking to consuming fish as it is a rich

source of omega three fatty acids, which helps in keeping the mind and body fit.

- Aquaculture or fish farming is the highest sector in the graph of food production, followed by normal fishing. In the year 2006, aquaculture comprised 47 percent of the total food production. That number has been on a steady rise and is said to rise further owing to the popularity of the practice.

- It has been widely observed that where fish farming is practiced, it is done in a significantly small scale. Sometimes, farmers tend to over breed and overpopulate the fish in severely congested and small areas. They will not know when to stop and will have several fish sharing a small space.

- The practice of overcrowding leads to pollution as thorough cleaning is not possible and because fish eat up their food from the same polluted water inside the tanks. Therefore, the possibility of the fish getting ill is higher, which in turn is detrimental to the health of individuals who consume them.

- The vicious cycle does not end here. In the efforts to treat the pollution as well as the ill fish, the water tank is further pumped in with purifying vaccines, antibiotics and chemicals for which government-prohibited substances are utilized. Despite being prohibited (but given the lack of quality checks and security), these fish find their way to our stomachs. So be very careful about it and make sure that the tank is absolutely clean and meets the standard clean tank norms in order to prevent any contamination. If you feel that your tank is overcrowding, then expand it or shift some of the fish to another tank.

- When the fish from the fish farms escape, it obviously leads to not just the financial loss to fish farmers. It also negatively contributes to the ecosystem. So be very careful with it and make sure that all the security features are in check.

- Most people select Tilapia for fish farming since it can be grown in tanks, rather than requiring pools. Salmon, cod, carp, and catfish are some of the other common types of fish that can be farmed. Each type of fish will have an advantage and a disadvantage, and it will be up to you to choose the one that you think will suit your needs. Remember, if something is working for a friend or family, then it might not work for you. You have to choose depending on the conditions that prevail in your locality. One tip is to look at your neighbor's farm and look at what is working best for them. You can also adopt the same species of fish and maybe change the variety.

- If you choose to grow salmon in your fish farm, then you must be aware that they are very high maintenance, as they need almost about five times as much food in proportion to their own body weight. If you underfeed them, then you might end up with sick fish and worse, cannibals! Not to say that you mustn't consider farming salmon but just understand that you will have to take more care.

- Shrimp is one of the most destructive to farm. This is because the salinity of the soil tends to increase when growing shrimps. The trend among farmers of changing farming locations often leaves the place unused and unfit for agricultural purposes after. So you must thoroughly reconsider farming shrimp as not only might you end up having a bad experience with them but also have a water source that is not clean enough to continue farming.

- One crucial tip for backyard fish farmers is that before you begin buying your tanks, fish, and other supplies, you must take care of the sewage management and water purification processes first. Do some research on the various methods and techniques, for instance, raceways, recirculating, or cages that will provide you with convenience and at the same time, suit your particular needs. Remember that you must keep the ecology in mind when you decide to take up fish farming and must ensure that you are not causing the environment any harm.

# Basic Aquaculture Tools and Equipment for Small Farmers

There are so many ways to put up a fish farm. You may choose to cultivate fish in a man-made or natural outdoor body of water, such as river or lake (by placing barriers between the other aquatic life and your farmed fish). Alternatively, you may set up above-ground tanks and filters that are established away from a natural water source.

The following are some of the specific tools and equipment that you need to put up a small fish farm:

## Filter and Tanks

For fish farms that are to be put up above-ground where you will not use a natural source of water, make use of large-sized tanks or pools to keep your fish. For these types of systems, aeration is particularly crucial as there will be no natural circulation by which the oxygenated water can enter into the farming area. You will also require filters that will keep the water clean as different concentrations of nitrites, and ammonia (the by-product from your fish) flow through.

You must buy yourself good quality ones, and the best thing is to buy them online. This will allow you to look for discounts, and you will also be able to conduct research on the company that you are buying from. You must have a lot of spares at all times as you never know when you might have to change the existing system. You cannot take a chance with your fish as they are perishable and just a matter of seconds is enough for a disaster to ensue.

## Diffusers and Aerators

Aerators may either be floated on a movable raised area in order to move across the water surface or be kept stationary, moving oxygen into the water surface through attached tubing. A diffuser is composed of porous media like rubber, wood or stone. As air flows through the rubber tubing into the diffuser, the medium will release it as small bubbles, which tend to dissolve more easily in the water.

Unlike aquariums, you do not have to buy fancy ones as it is more about purpose than beauty. Basic ones will work fine but make sure that they are of high quality. You do not want to take chances with cheap materials as they might start to pollute your tank. Some people do not pay attention to the small things, and this can be quite a bad thing. You have to be sure of the quality of material that you wish to use, and this will go a long way in ensuring that you provide your fish with a highly suitable environment.

## Submersible Nets and Cages

For an outdoor fish farm, separate other fish from your farmed fish in the river or lake with the use of submersible cage or netting. Cages may be made of any robust material like metal or plastic. They're essentially perforated or mesh walls that permit water to flow through while keeping unwanted fish from getting in and disturbing with your livestock.

Submersible netting provides the same effect, although they provide more flexibility. Nets are collapsible for easy transport and storage outside of the water. Again, it is important for you to concentrate on the quality. If you have a poor quality net, then the fish might get entangles in it, and that can cause them to get injured. It might also contaminate the water, which can be equally bad. One trick is to talk to the local farmers and find out about the brand that they use for their catch. This will allow you to buy sturdy ones for your farm.

## Docks and Boats

For an outdoor fish farm, you may have to transfer your captive fish in order to harvest, monitor or feed them. A movable or modular floating dock will permit access to your fish while giving sufficient surface area to hold nets, feed as well as other equipment. If you have a limited amount of fish farming equipment, or if you have to cover a big area more rapidly, use a motorized boat to transport your fish farm. Any kind of motorized boat will do the work, such as aluminum fishing boats or durable inflatable rafts, provided that it has sufficient room for all of your stuff. You can also decide to buy a used boat in order to cut down on your costs. You can borrow one every now and then if you think that buying a boat for yourself will turn out to be an expensive affair.

## Filtration

Water filtration is crucial in getting rid of fecal material, as well as other organic wastes. This factor is of significant concern to prevent illnesses from decimating the whole stock, particularly in a tank containing of hundreds of fishes. Not only is solid matter eliminated but contaminants such as ammonia and nitrogen should be gotten rid of. Ammonia is toxic while nitrogen will boost the growth of algae. Fish will release around 14 grams of ammonia for every pound of feed they consume.

Water filtration systems need their own equipment including a variety of filter media, pumps and settling tanks, depending on which specific type of method is employed.

## Decorations

Although not necessary, you can buy decorations for your tank. This can include pebbles, rocks, colored stones, fake plants, real plants, colored gravel, etc. it is completely up to you to choose the decorations but make sure that they are pool safe. There can be some plants that might not be suited for the aquarium and might only cause pollution, you have to identify these and not have them in your pool.

## Costs

Aeration and pumping are the two biggest consumers of electricity in every aquaculture system. In operating any type of aquaculture system, every aquaculturist must take into consideration the costs that shall be incurred to cover expenses for electricity, labor, and feeds. Of course, you'll also have to take the cost of seed stock or fingerlings into account. While some species do come at lower prices, availability also affects value. Don't just settle for the cheapest fish though, as you'll have to consider consumption requirements and growth potential as well. Simply put, you shouldn't be hasty when choosing among different fish species, as making a mistake could prove to be costly.

# Expansion for Your Fish Farm

Whether you have a backyard farm or a cage system, there will come a time when you will want to expand your farm. Expansion is a great idea as you will be able to make the most of your business and once you develop the expertise in the field of fish farming, you will have an opportunity to put it to good use. But before you decide to expand, you must consider the following aspects, in order to do all the right things for your farm.

## Space

The first thing to consider will be the space in your backyard. You must have enough space to expand the pool considerably as you will have to add a substantial amount of fish in order to have an expansion. So start by looking at the amount of space that you have. Here, you can either decide to have another pool next to the existing one or can simply expand the existing one. You can also have another one placed in a different location if there is spare place around your house. You can also utilize your terrace for this purpose, but you may have to add in a roof to protect your fish from the harsh rays of the sun. You can expand your cage system by buying yourself a bigger cage and your aquaponics by investing in bigger pools. As for your raceway, you can ask for permission and expand the tunnel.

## Budget

The next thing that you must consider is the budget. You cannot have an expansion plan with an unlimited budget. That would be foolish. You will have to sketch out a budget, just like you did when starting out. Generally, people invest a small portion of their profit in order to expand. You can also decide to avail a loan for the expansion as you will be able to repay it if your expansion goes well. You can consult with a family member or your partner and decide on how much you wish to invest for your expansion plans. Just like with the initial plan, you must set a budget that is lesser than what you think it will cost you and this will help you to not spend more than necessary.

## Permission

The next step is to avail permission from the authorities. Remember that expanding will mean you will use some more of the ocean, lake; river, reservoir, etc. and this will have to be approved. You don't want to get into any trouble later and so, it is important that you take due permission from the authorities. Sometimes, you might have to display patience as they might approve it easily. You might also have to do a few rounds to the authority's office, but it will only work well for you in the end.

## Variety

When you expand, you will be able to introduce a lot more variety. This will only make your business even more profitable. You must understand that the bigger the variety, the more the customers. You might be able to grow both edible and ornamental fish, and this will only increase your business scope. But remember that you cannot club all the varieties together as the fish might turn on each other. You might have to buy different ponds for each type and so, it is best that you decide on just a few varieties more.

## Corrections

When you decide to expand, you will have the option to correct any mistakes in your existing plans. It will give you the opportunity to make amends and fix any previous flaws. You might not have to spend extra for it and might get to use the labor to help fix any of the old problems. For this, you might have to make a list of everything that you have to fix as it will be easy for you to fix it.

## Security

Once you expand your farm, you might have to increase the security feature in your house. There might be the danger of theft and so, it is important that you make use of fences, barbed wires, cameras, etc. you must also keep an eye on the customers that come along as they might be up to some mischief.

## Marketing

Once you expand, you will have to get word of your business out. For this purpose, you will have to make use of advertising and other such medium and more on this is explained in the next chapter.

# Marketing

Once you decide to expand and do expand your aquaculture business, it is obvious that you will be able to produce more fish, and it will be important to get your word out. But marketing is never easy, and there are many things that you must do in order to market your produce. In this chapter, we will look at the various things that you must consider in order to have a successful marketing strategy in place for your business. You will move from having mere customers to having a large customer base for your produce.

## Market area

The very first thing that you must consider is the market area. This means you must understand the various markets that you have for your fish. You must do a thorough research for this and might have to employ a team who will help scour the various markets for you. You will already have a few customers, and you can build on these. You must look at your competitors' markets as well as you will be able to find niche markets for your type of fish. Remember that it should go much beyond mere internet searches and you must do on-site inspections and schedule visits to the markets in order to identify your potential market areas. It is believed that most markets located within a ten-mile radius of your farm will be your best bet.

## Market segments

The term "market" is quite a vast one and incorporates buyers, sellers and also middlemen. But you have only to look for your potential buyers in the market and so, once you decide on the markets, you must conduct a thorough research to identify all the potential customers. Say, for example, there are ten buyers in the market, all 10 of these cannot be seen as your potential buyers, and you must assess as to how many of these can be your potential customers. This might sound complex, but it is easier than you think it is.

All you have to do is employ observational techniques to assess which customer is purchasing what type of fish. This will give you a fair idea of the number of people who buy your type of fish and so, you will be able to surmise your potential customers. Remember that most customers for aquaculture are supermarkets, hotels, restaurants, wholesalers, etc. so it is best that you observe their dealers closely.

## Target buyers

Once you identify your potential buyers, you must narrow them down further and decide on your target buyers. This will be easy to do as you will be able to group them into individual groups. So say you have ten supermarkets and ten wholesalers, you will be able to narrow these down to 5 supermarkets and five wholesalers and have ten target buyers in your kitty. Similarly, you must identify your set of target buyers and come up with potential and rigid numbers. Remember that these are just examples and if you deal in ornamental fish such as the goldfish, then your target audience will be pet stores, fish enthusiasts, etc. Your target audience will vary, but you have to be able to come up with a proper customer base.

## Needs and potential

Once you identify your target buyers, you must study each one in detail in order to identify their needs. This means that you must conduct a market survey and understand what the buyers need. This will help you skim the best buyers from the rest and will be able to cater to them in a better manner. The same goes for understanding their potential. You will be able to know how much the particular buyer will buy from you and so, you will have a chance to prepare for the supply. Remember that customer needs are volatile and so, you must be prepared to cater to their needs in order to hold on to your customer base.

## Costs

Understand that you will have to set a separate budget for your marketing. Marketing is not a cheap activity, and you will have to shell out quite a bit, in order to find potential customers. You must decide on the cost depending on your needs and have a certain amount set aside for it. Remember, if you do it right, then you will be able to make back a lot more than what you will invest in your marketing. However, if something goes wrong, then you might suffer a setback. So the best thing to do is to try and make back whatever sum that you invest so that you will not lose any money, and any surplus that you make will be your profit.

## Advertising

Advertising refers to having your business printed on pamphlets and having radio jingles mention about your business. How you choose to advertise is completely up to you and you must make sure that your target buyers get to hear a word about your business. This will help you get your potential buyers as also get other buyers to come your way. Most types of advertising can be expensive, but it will greatly benefit your business to have your ads getting noticed by a large population. You can get the help of a publicist and try and get the word out about your business to reach as many people as possible.

## Online presence

The internet is your best friend and how well you use it will determine how much you will get noticed. There are many things that you can do on the internet to connect with your customer base and establish a clear line of communication with them. You can be present on any of the social media sites and be able to speak with your customers. You will be able to announce any new variety of fish and also be able to answer customer queries. Being able to announce new things will give your business a big boost and having a way to reach out to more and more people will only help your business grow bigger. You do not have to invest money to open a Facebook page and write about your business. All that it will take is a few minutes time, and you will be able to broadcast about your business in no time.

## Delivery systems

Once you have everything sorted, you must decide on the delivery systems. This means understanding how you will deliver the products and the mode of transport. You might be used to having customers pick up the fish from your place but once you decide to expand and go bigger, you will have to cater to a larger audience base and might have to deliver more than just 10 to 20 fish and that number might run in the hundreds. You will have to tie up with specialized courier services, who will help transport live produce. You can also decide to have your own delivery systems in place as it will be easy for you to do many in a day. The choice will be completely up to you.

## Finances

Once you have your business up and running thanks to your marketing endeavors, you will be able to make profits. You must have a plan in place to do whatever is right with your finances and be in a position to make the best use of it. If you need help with that, then you can employ someone to take care of the finances. Remember that you cannot do all of it by yourself and you will need help from personnel as and when your business starts to expand. You will have to employ specific people to help you out. This is explained in detail in the next chapter.

## After sales care

Remember that your relationship with your customer is not over when you deliver the fish to them. You will have to maintain a consistent relationship in order for them to order more from you. For this, you will have to provide them with an effective after sales service. This means that you must keep in touch with them, and if they have a grievance, then you must address it as fast and efficiently as you can. Many times, there can be some dead fish in the batch and it is important that you replace these in order to maintain better customer relationships.

## Employee recruitment

Once you expand and start having a good business, it is obvious that you will need some help with the store. For this, you will have to advertise the jobs that you wish to offer and here are things that you can do.

### *Departments*

It is obvious that you will need personnel for different departments in your business. You must decide on the various job posts and also decide on how much you will pay to each one. You will need a manager, someone to remove the fish, someone to transport, someone to look into the finances, etc. you will have to decide on the types of jobs that you will offer as also the number of people that you will have to hire for your business.

## *Recruiting*

The next step is to recruit the personnel. For this purpose, you can post ads on the internet and look for people that might be interested in taking up the job. You can post the ads on social networking sites or have them circulate amongst friends and family. Once you have people coming in, you might have to interview them. Remember that it will be necessary for each one to have some basic knowledge about fish farms and they will also have to demonstrate their skills so that you will get a chance to hire the right people.

## *Classes*

The next thing is to train the individuals. This will require you to teach them the ways of your farm as everybody will have a different work ethic. You must teach them the right way to handle fish, managing the clients, keeping the pools clean, etc. Once you are done, you might have to see them do it in order to be convinced of their skills.

## *Online classes*

If you do not have the time to teach one what they have to do, then you can make them watch online videos. This will give them a good idea, and if they are already well versed with it, then you might only have to brief them.

## *Assist*

Sometimes, you might have to assist them. Employing someone will not mean that you can take a back seat, and you will have to continue taking care of the business. You have to keep an eye on your workforce and make sure that everything is going smoothly. If any of them need assistance, then you have to be available to provide it.

# Conclusion

Thank you again for downloading this book!

I hope this book was able to give you a better understanding of aquaculture. As is apparent, there will be a few challenges to face but it is important to keep the faith and keep moving, in order to make it big in the business. It will seem a bit overwhelming at first, but that should not deter you from continuing with fish farming.

The next step is to finally decide whether you should engage in aquaculture endeavors. Once you've made your decision and assuming that you're already raring to start farming fish, simply follow the steps presented in this book as well as the practical tips in order to be successful in your aquaculture venture.

Finally, if you enjoyed this book, please take the time to share your thoughts and post a review on Amazon. It'd be greatly appreciated!

Thank you and good luck!

Kenn

# Photo Credits

Aquaculture by Bytemarks
https://www.flickr.com/photos/bytemarks/5210691165
https://creativecommons.org/licenses/by/2.0/legalcode

Untitled by Capecodprof
http://pixabay.com/en/trout-fish-aquarium-seafood-fresh-233571/
https://creativecommons.org/licenses/by/2.0/legalcode

Aquaculture_tank_design_2nd_edition by Danieljackson
http://commons.wikimedia.org/wiki/File:Aquaculture_tank_design_2nd_edition.JPG
Public Domain

November's Bytemarks Lunch visit to the McKinley aquaculture fish farm.by bytemarks
https://www.flickr.com/photos/bytemarks/5211291780
https://creativecommons.org/licenses/by/2.0/legalcode

Conclusion
November's Bytemarks Lunch visit to the McKinley aquaculture fish farm.by bytemarks
https://www.flickr.com/photos/bytemarks/5210687851
https://creativecommons.org/licenses/by/2.0/legalcode

# Other Books by Kenn Christenson

Below you'll find my other books that are available on Amazon and Kindle as well. Click on the links below to check them out. You can visit my author page on Amazon to see other work done by me.

Goatherd 101: An Introduction To Raising Meat and Dairy Goats

http://www.amazon.com/dp/B00N9K5J5Y

Chickens!: The Best Backyard Chicken Breeds for Organic Meat and Eggs

http://www.amazon.com/dp/B00R0IMA6M

You can simply search for these titles on the Amazon website to find them or type in the links that are below the book titles.

Printed in Great Britain
by Amazon.co.uk, Ltd.,
Marston Gate.